For Someone Special

To Discover The Mystery

A Book By M. Areeb

The Fabric Of Space & Time

By Muhammad Areeb

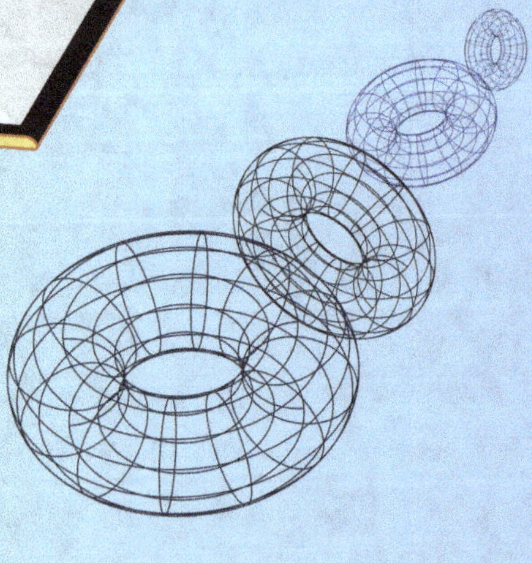

Introduction

This book explains Spacetime which is the fabric of the universe. It is the arena in which all events take place, and it is the medium through which all forms of energy and matter propagate. Spacetime is curved by mass and energy, and this curvature is what gives rise to gravity.

The study of spacetime has a long and rich history. The first person to propose the concept of spacetime was Gottfried Leibniz in the 17th century. However, it was Albert Einstein who first developed a comprehensive theory of spacetime, general relativity. Einstein's theory showed that spacetime is not a fixed background, but is instead dynamic and curved by mass and energy.

Spacetime is a complex and mysterious place. There are many things about spacetime that we do not yet understand, such as the nature of black holes, the possibility of time travel, and the origin of the universe. These mysteries continue to challenge and fascinate physicists and philosophers alike.

Our understanding of spacetime is constantly evolving. As we learn more about the universe, we gain a better understanding of how spacetime works. In the future, we may be able to develop new technologies that allow us to travel through spacetime, or even to create new spacetimes. The possibilities are endless. This book explores the fabric of spacetime, from its origins to its future. We will discuss the history of spacetime, the different theories of spacetime, and the mysteries that still surround spacetime. We will also explore the implications of spacetime for our understanding of the universe and our future as a species. The Fabric of Space & Time is a journey through the most fundamental fabric of the universe. It is a journey that will challenge your understanding of reality and leave you wondering what else is out there.

Spacetime is more than just a background for events. It is a dynamic and active participant in the universe. It is the fabric of reality, and it is what gives the universe its structure and meaning. The study of spacetime is one of the most important and exciting areas of physics today. As we learn more about spacetime, we are learning more about the universe itself. And as we learn more about the universe, we are learning more about ourselves.

The Fabric of Space & Time is a journey of discovery. It is a journey that will take you to the very heart of the universe. It is a journey that will change the way you see the world.

****

About The Author

This book has been written by Muhammad Areeb, a writer, an entrepreneur, a computer programmer, physics and a youtuber who is passionate about doing research in the field of science especially astronomy and physics. The author has a deep understanding in the field of astronomy and physics.

****

From The Author

"I am writing this book to share my knowledge of spacetime with a wider audience. I believe that spacetime is one of the most important concepts in physics, and I want to help people to understand it better.

I will be writing about spacetime in a clear and concise way. I will use simple language that is easy to understand, but I will also be accurate and rigorous. I will cite my sources so that readers can verify my claims.

I hope that this book will help people to understand spacetime better. I hope that it will inspire people to learn more about physics and mathematics."

****

What You Will Learn

What Is Space & Time?

Spacetime is the fabric of the universe. It is the arena in which all events take place, and it is the medium through which all forms of energy and matter propagate. Spacetime is curved by mass and energy, and this curvature is what gives rise to gravity.

****

INTRODUCTION

Spacetime is the fabric of the universe. It is the four-dimensional continuum that encompasses all of space and time. In spacetime, everything that exists, from planets and stars to atoms and subatomic particles, is interconnected. The concept of spacetime was first developed by Albert Einstein in his theory of general relativity. Einstein showed that space and time are not separate entities, but rather two aspects of the same thing. The curvature of spacetime is caused by the presence of mass and energy. The more mass and energy an object has, the more it curves spacetime. This curvature is what causes gravity.

THE NATURE OF SPACETIME

Spacetime is a mathematical construct, but it is also a physical reality. It is the stage on which the entire universe unfolds. Spacetime is not a static thing, but rather it is constantly evolving. The curvature of spacetime is constantly changing, as the universe expands and galaxies move apart.

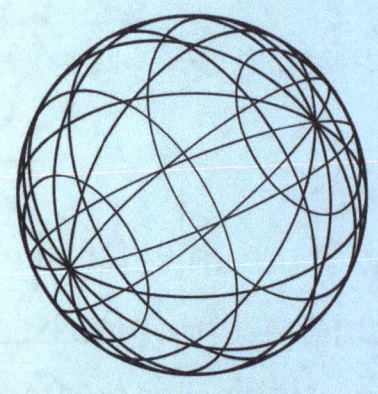

THE EFFECTS OF SPACETIME CURVATURE

The curvature of spacetime has a number of effects on objects in the universe. For example, the curvature of spacetime can cause objects to follow curved paths, such as the orbits of planets around stars. The curvature of spacetime can also cause objects to accelerate, such as the acceleration of objects toward black holes.

The curvature of spacetime can also have a profound impact on the passage of time.

FOR EXAMPLE

Time passes more slowly near a massive object, such as a black hole. This is known as gravitational time dilation.

THE FUTURE OF SPACETIME

The future of spacetime is uncertain. However, we know that the curvature of spacetime is constantly evolving, as the universe expands and galaxies move apart. This evolution of spacetime could have a profound impact on the future of the universe.

FOR EXAMPLE

If the universe continues to expand at an accelerating rate, then the curvature of spacetime will eventually become so great that even light will not be able to escape. This would lead to a "big rip," in which the universe would be torn apart.

On the other hand, if the expansion of the universe eventually slows down and reverses, then the curvature of spacetime will eventually become so negative that the universe will collapse in on itself. This would lead to a "big crunch," in which the universe would be reborn in a new Big Bang.

4

CONCLUSION

Spacetime is a fundamental concept in physics. It is the fabric of the universe, and it is what gives rise to gravity. The curvature of spacetime is a complex phenomenon, but it is essential to understanding the universe.

The future of spacetime is uncertain, but it is likely to be a fascinating journey. We are only just beginning to understand the nature of spacetime, and there is much that we still do not know. However, as we continue to explore the universe, we will learn more about this fundamental aspect of reality.

****

INTRODUCTION

The dimensions of spacetime are the four directions in which we can move: up-down, left-right, forward-backward, and time. These four dimensions are interconnected, and they all play a role in the way the universe works.

THE DIMENSIONS OF SPACE

The three dimensions of space are the directions in which we can move through space. We can move up and down, left and right, and forward and backward. These three dimensions are what allow us to navigate the world around us.

The three dimensions of space are not completely independent of each other. For example, if we move forward in time, we also move forward in the other two dimensions of space. This is because the three dimensions of space are all part of the same spacetime continuum.

THE DIMENSION OF TIME

The fourth dimension is time. Time is the direction in which we can move through time. We can move from the past to the present and to the future. Time is what allows us to experience change and measure the passage of events.

Time is not completely independent of the other three dimensions of space. For example, the more massive an object is, the slower time passes near that object. This is known as gravitational time dilation.

THE INTERACTIONS OF THE DIMENSIONS

The dimensions of spacetime interact with each other in a number of ways. For example, the curvature of spacetime can affect the passage of time. This is known as gravitational time dilation.

Another example of the interaction of the dimensions is the way that mass and energy warp spacetime. This warping of spacetime is what causes gravity.

The warping of spacetime is not uniform. The more mass and energy an object has, the more it warps spacetime. This is why objects with mass, such as planets and stars, have a gravitational pull.

THE FUTURE OF THE DIMENSIONS

The future of the dimensions is uncertain. However, we know that the dimensions are constantly evolving. For example, the expansion of the universe is causing the dimensions of space to stretch. This stretching of the dimensions could have a profound impact on the future of the universe.

FOR EXAMPLE

If the expansion of the universe continues at an accelerating rate, then the dimensions of space could eventually become so stretched that they would tear apart. This would lead to a "big rip," in which the universe would be torn apart.

On the other hand, if the expansion of the universe eventually slows down and reverses, then the dimensions of space could eventually become so compressed that they would collapse in on themselves. This would lead to a "big crunch," in which the universe would collapse in on itself.

CONCLUSION

The dimensions of spacetime are a fundamental aspect of the universe. They are what allow us to experience the world around us, and they are what give rise to gravity. The interactions of the dimensions are complex, but they are essential to understanding the universe.

The future of the dimensions is uncertain, but it is likely to be a fascinating journey. We are only just beginning to understand the nature of the dimensions, and there is much that we still do not know. However, as we continue to explore the universe, we will learn more about this fundamental aspect of reality.

****

Curvature Of Spacetime By Mass & Energy

INTRODUCTION

The curvature of spacetime is a fundamental concept in physics. It is how mass and energy warp the fabric of the universe. This warping of spacetime is what causes gravity.

THE THEORY OF GENERAL RELATIVITY

Albert Einstein developed the theory of general relativity in the early 20th century. Einstein's theory showed that mass and energy are equivalent and that they both warp spacetime. The more mass and energy an object has, the more it warps spacetime.

THE CURVATURE OF SPACETIME

The curvature of spacetime can be visualized by thinking of a trampoline. If you place a heavy object on a trampoline, the trampoline will bend down. The more massive the object, the more the trampoline will bend. This is similar to the way that mass and energy warp spacetime.

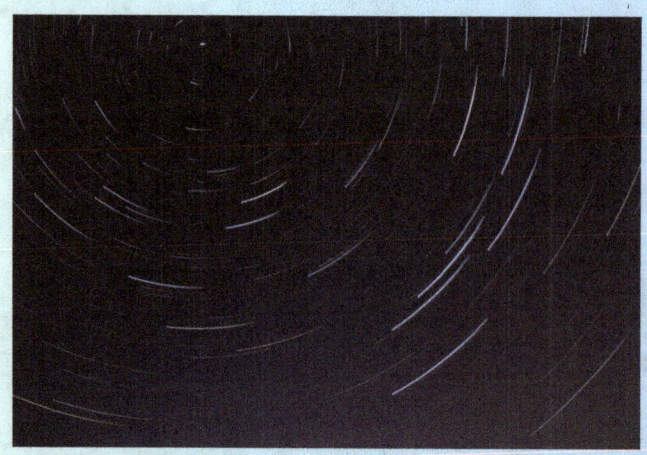

THE EFFECTS OF CURVATURE

The curvature of spacetime has a number of effects on objects in the universe. For example, the curvature of spacetime can cause objects to follow curved paths, such as the orbits of planets around stars. The curvature of spacetime can also cause objects to accelerate, such as the acceleration of objects toward black holes.

GRAVITATIONAL TIME DILATION

Another effect of the curvature of spacetime is gravitational time dilation. This is the phenomenon where time passes more slowly near a massive object, such as a black hole. This is because the curvature of spacetime near a massive object distorts the flow of time.

11

THE FUTURE OF THE CURVATURE OF SPACETIME

The future of the curvature of spacetime is uncertain. However, we know that the curvature of spacetime is constantly evolving. For example, the expansion of the universe is causing the curvature of spacetime to decrease. This could have a profound impact on the future of the universe.

FOR EXAMPLE

if the expansion of the universe continues at an accelerating rate, then the curvature of spacetime could eventually become so small that gravity would no longer exist. This would lead to a universe in which objects would no longer be attracted to each other, and the universe would eventually become a cold and dark place.

On the other hand, if the expansion of the universe eventually slows down and reverses, then the curvature of spacetime could eventually become so large that gravity would become infinitely strong. This would lead to a universe in which all objects would be pulled together into a single point, and the universe would eventually collapse in on itself.

CONCLUSION

The curvature of spacetime is a complex phenomenon, but it is essential to understanding the universe. The future of the curvature of spacetime is uncertain, but it is likely to be a fascinating journey. We are only just beginning to understand the nature of the curvature of spacetime, and there is much that we still do not know. However, as we continue to explore the universe, we will learn more about this fundamental aspect of reality.

****

The History Of Spacetime

The study of spacetime has a long and rich history. The first person to propose the concept of spacetime was Gottfried Leibniz in the 17th century. However, it was Albert Einstein who first developed a comprehensive theory of spacetime, general relativity. Einstein's theory showed that spacetime is not a fixed background, but is instead dynamic and curved by mass and energy.

****

The Study Of Spacetime
From
Ancient Times To The Present Day

INTRODUCTION

The study of spacetime has a long and rich history. From ancient philosophers to modern physicists, people have been trying to understand the nature of space and time.

ANCIENT CONCEPTIONS OF SPACE AND TIME

In ancient Greece, philosophers such as Aristotle and Plato developed theories of space and time. Aristotle believed that space and time were absolute, meaning that they existed independently of objects and events. Plato, on the other hand, believed that space and time were relative, meaning that they were dependent on objects and events.

THE DEVELOPMENT OF MODERN PHYSICS

The development of modern physics in the 17th and 18th centuries led to a new understanding of space and time. Isaac Newton's theory of gravity showed that space and time were not absolute, but rather they were relative to the observer. This led to the development of the theory of relativity, which is the modern understanding of space and time.

THE THEORY OF GENERAL RELATIVITY

Albert Einstein developed the theory of general relativity in the early 20th century. Einstein's theory showed that space and time are not separate entities, but rather they are two aspects of the same thing. This thing is called spacetime.

The theory of general relativity also showed that the curvature of spacetime is caused by the presence of mass and energy. The more mass and energy an object has, the more it curves spacetime. This curvature is what causes gravity.

THE STUDY OF SPACETIME TODAY

The study of spacetime is a rapidly evolving field. Today, physicists are using cutting-edge technology to study the nature of spacetime. They are also using mathematical models to try to understand how spacetime works.

The study of spacetime is a fascinating and challenging field. It is a field that is constantly evolving, and it is a field that is likely to yield many new insights in the years to come.

****

Key Figures In The Development Of Our Understanding Of Spacetime

INTRODUCTION

The study of spacetime has a long and rich history, and there have been many key figures who have contributed to our understanding of this fundamental concept.

ANCIENT PHILOSOPHERS

Some of the earliest ideas about space and time can be found in the writings of ancient philosophers. In ancient Greece, for example, Aristotle believed that space and time were absolute, meaning that they existed independently of objects and events. Plato, on the other hand, believed that space and time were relative, meaning that they were dependent on objects and events.

GALILEO GALILEI

In the 17th century, Galileo Galilei conducted a series of experiments that showed that space and time were not absolute. For example, he showed that the speed of light was constant, regardless of the motion of the observer. This showed that space and time were not independent of objects and events, as Aristotle had believed.

ISAAC NEWTON

In the 18th century, Isaac Newton developed a theory of gravity that showed that space and time were relative to the observer. Newton's theory showed that the motion of objects was affected by their mass and the curvature of spacetime. This was a major breakthrough in our understanding of space and time.

ALBERT EINSTEIN

In the early 20th century, Albert Einstein developed the theory of general relativity. Einstein's theory showed that space and time are not separate entities, but rather they are two aspects of the same thing. This thing is called spacetime.

Einstein's theory also showed that the curvature of spacetime is caused by the presence of mass and energy. The more mass and energy an object has, the more it curves spacetime. This curvature is what causes gravity.

OTHER KEY FIGURES

In addition to the figures mentioned above, there have been many other key figures who have contributed to our understanding of spacetime. These include Hermann Minkowski, who developed the mathematical formulation of spacetime; Karl Schwarzschild, who solved Einstein's field equations for a spherically symmetric mass distribution; and Stephen Hawking, who made major contributions to our understanding of black holes and the Big Bang.

THE FUTURE OF OUR UNDERSTANDING OF SPACETIME

Our understanding of spacetime is still evolving, and there are many unanswered questions. For example, we do not yet fully understand how spacetime works at the quantum level. We also do not know what happens to spacetime at the beginning and end of the universe.

However, the study of spacetime is a rapidly evolving field, and new insights are being made all the time. As we continue to study spacetime, we are likely to learn more about the fundamental nature of the universe.

****

The Evolution Of Our Understanding Of Spacetime

INTRODUCTION

Our understanding of spacetime has evolved over time, as new discoveries have been made and new theories have been developed.

EARLY IDEAS

Some of the earliest ideas about space and time can be found in the writings of ancient philosophers. In ancient Greece, for example, Aristotle believed that space and time were absolute, meaning that they existed independently of objects and events. Plato, on the other hand, believed that space and time were relative, meaning that they were dependent on objects and events.

THE RISE OF RELATIVITY

In the 17th century, Galileo Galilei conducted a series of experiments that showed that space and time were not absolute. For example, he showed that the speed of light was constant, regardless of the motion of the observer. This showed that space and time were not independent of objects and events, as Aristotle had believed.

In the 18th century, Isaac Newton developed a theory of gravity that showed that space and time were relative to the observer. Newton's theory showed that the motion of objects was affected by their mass and the curvature of spacetime. This was a major breakthrough in our understanding of space and time.

THE THEORY OF GENERAL RELATIVITY

In the early 20th century, Albert Einstein developed the theory of general relativity. Einstein's theory showed that space and time are not separate entities, but rather they are two aspects of the same thing. This thing is called spacetime.

Einstein's theory also showed that the curvature of spacetime is caused by the presence of mass and energy. The more mass and energy an object has, the more it curves spacetime. This curvature is what causes gravity.

THE DEVELOPMENT OF QUANTUM GRAVITY

In the late 20th century, physicists began to develop a theory of quantum gravity. Quantum gravity is a theory that attempts to unify the laws of general relativity with the laws of quantum mechanics.

Quantum mechanics is a theory that describes the behavior of matter and energy at the atomic and subatomic levels. General relativity is a theory that describes the behavior of gravity on a large scale. The development of quantum gravity is a major challenge, but it is an important step in our understanding of the universe.

THE FUTURE OF OUR UNDERSTANDING OF SPACETIME

Our understanding of spacetime is still evolving, and there are many unanswered questions. For example, we do not yet fully understand how spacetime works at the quantum level. We also do not know what happens to spacetime at the beginning and end of the universe.

However, the study of spacetime is a rapidly evolving field, and new insights are being made all the time. As we continue to study spacetime, we are likely to learn more about the fundamental nature of the universe.

****

The Mysteries Of Spacetime

Spacetime is a complex and mysterious place. There are many things about spacetime that we do not yet understand, such as the nature of black holes, the possibility of time travel, and the origin of the universe. These mysteries continue to challenge and fascinate physicists and philosophers alike.

****

The Nature Of Black Holes & The Big Bang

INTRODUCTION

Black holes are regions of spacetime where gravity is so strong that nothing, not even light, can escape. They are formed when massive stars collapse at the end of their lives.

The boundary of a black hole is called the event horizon. Once something passes the event horizon, it is forever lost to the rest of the universe.

Black holes are one of the most mysterious objects in the universe. We do not fully understand how they work, and there are many unanswered questions about them.

THE BIG BANG

The Big Bang is the theory that the universe began with a very hot, dense state and has been expanding and cooling ever since. The Big Bang is supported by a wide range of evidence, including cosmic microwave background radiation, the abundance of light elements in the universe, and the redshift of distant galaxies.

The Big Bang is a hotly debated topic, and there are many unanswered questions about it.

FOR EXAMPLE

We do not know what caused the Big Bang, or what happened before the Big Bang.

However, the Big Bang is the best explanation we have for the origin of the universe, and it is a cornerstone of modern cosmology.

THE CONNECTION BETWEEN BLACK HOLES AND THE BIG BANG

There is a connection between black holes and the Big Bang. Black holes are thought to be the end product of the collapse of massive stars, and the Big Bang was the beginning of the universe. In a sense, black holes are like time capsules. They contain information about the universe as it was when the stars that formed them collapsed. By studying black holes, we can learn more about the early universe.

THE FUTURE OF OUR UNDERSTANDING OF BLACK HOLES AND THE BIG BANG

Our understanding of black holes and the Big Bang is still evolving, and there are many unanswered questions. However, the study of these objects is a rapidly evolving field, and new insights are being made all the time.

As we continue to study black holes and the Big Bang, we are likely to learn more about the fundamental nature of the universe.

****

INTRODUCTION

The possibility of time travel has been a topic of speculation for centuries. There are many different theories about how time travel might be possible, but there is no scientific consensus on the matter. One of the most famous theories of time travel is the "wormhole" theory. Wormholes are hypothetical tunnels that connect different points in spacetime. If wormholes exist, it might be possible to travel through them to different points in time.

Another theory of time travel is the "closed timelike curve" theory. Closed timelike curves are hypothetical paths through spacetime that allow an object to travel back in time. Closed timelike curves are very difficult to create, and it is not clear if they actually exist.

OTHER EXOTIC PHENOMENA

In addition to time travel, there are many other exotic phenomena that have been proposed by physicists and science fiction writers. These include teleportation, faster-than-light travel, & parallel universes.

TELEPORTATION

Teleportation is the hypothetical ability to transport an object or person from one location to another instantly. Faster-than-light travel is the hypothetical ability to travel faster than the speed of light. Parallel universes are hypothetical universes that exist alongside our own.

THE FUTURE OF EXOTIC PHENOMENA

The possibility of time travel and other exotic phenomena is still a matter of speculation. However, the study of these phenomena is a rapidly evolving field, and new insights are being made all the time. As we continue to study these phenomena, we are likely to learn more about the fundamental nature of spacetime. We may even one day discover that these phenomena are actually possible.

THEORETICAL AND EXPERIMENTAL CHALLENGES

There are a number of theoretical and experimental challenges that would need to be overcome in order to make time travel or other exotic phenomena a reality.

One challenge is that the laws of physics as we understand them seem to prohibit time travel.

FOR EXAMPLE

The speed of light is the ultimate speed limit in the universe, and it is not clear how anything could travel faster than light.

Another challenge is that there is no known way to create wormholes or closed timelike curves. These objects are hypothetical, and it is not clear if they actually exist.

Even if wormholes or closed timelike curves do exist, it is not clear if they would be stable. If they were not stable, they would collapse before anything could pass through them.

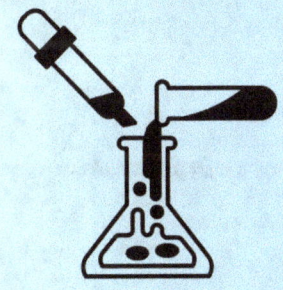

CONCLUSION

The possibility of time travel and other exotic phenomena is a fascinating topic. However, there are a number of theoretical and experimental challenges that would need to be overcome in order to make these phenomena a reality.

As we continue to study these phenomena, we are likely to learn more about the fundamental nature of spacetime. We may even one day discover that these phenomena are actually possible.

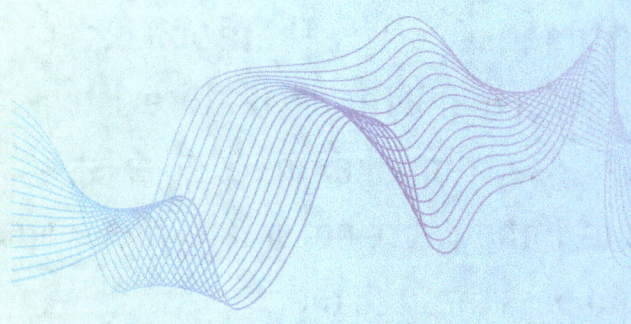

****

The Implications Of Spacetime For Our Understanding Of The Universe

INTRODUCTION

The concept of spacetime has profound implications for our understanding of the universe. It tells us that space and time are not separate entities, but rather they are two aspects of the same thing. This has a number of implications, including:

- Spacetime is curved by mass and energy. This curvature is what causes gravity. For example, the sun's mass curves spacetime, which is what causes the planets to orbit the sun.

- The speed of light is the ultimate speed limit in the universe. This means that nothing can travel faster than the speed of light, not even information. This has implications for our understanding of the universe, as it means that we can never travel back in time or visit other galaxies instantaneously.

- The universe is expanding. This expansion is accelerating, and it is not clear what will happen to the universe in the future. If the acceleration continues, the universe will eventually tear itself apart. However, if the acceleration slows down or stops, the universe may eventually collapse in on itself.

- Time travel and other exotic phenomena may be possible. However, there is no scientific consensus on this matter. Some physicists believe that time travel is theoretically possible, but that it would require the existence of wormholes or other exotic objects. Other physicists believe that time travel is impossible, as it would violate the laws of physics.

The concept of spacetime is still a relatively new one, and there is much that we do not yet understand about it. However, it is a fundamental concept in physics, and it has a profound impact on our understanding of the universe.

THE FUTURE OF SPACETIME

The study of spacetime is a rapidly evolving field, and new insights are being made all the time. As we continue to study spacetime, we are likely to learn more about the fundamental nature of the universe.

Here are some of the questions that physicists are still trying to answer about spacetime:

- What is the relationship between spacetime and quantum mechanics?
- What is the origin of spacetime?
- What is the ultimate fate of the universe?

These are just a few of the questions that physicists are still trying to answer about spacetime. As we continue to study spacetime, we are likely to learn more about the fundamental nature of the universe, and we may even one day discover the answers to these questions.

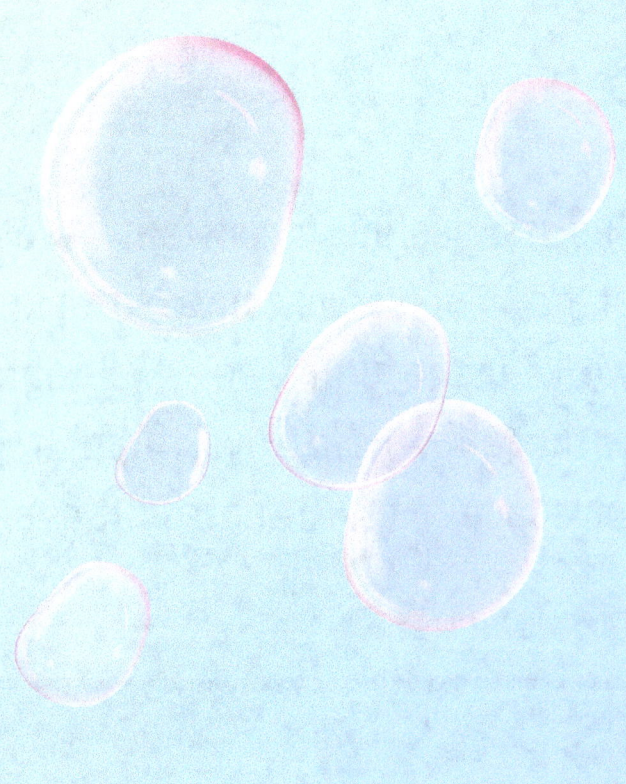

****

The Future of Spacetime

Our understanding of spacetime is constantly evolving. As we learn more about the universe, we gain a better understanding of how spacetime works. In the future, we may be able to develop new technologies that allow us to travel through spacetime, or even to create new spacetimes. The possibilities are endless.

****

WHAT DOES IT MEAN?

Our understanding of spacetime has evolved, and it is likely to continue to evolve in the future. As we learn more about the universe, we are likely to gain a deeper understanding of spacetime and its role in the cosmos.

One of the most important challenges facing physicists today is to reconcile the theory of general relativity, which describes gravity on a large scale, with quantum mechanics, which describes the behavior of matter and energy on a small scale. This is known as the "quantum gravity problem," and it is one of the most important unsolved problems in physics. If physicists can solve the quantum gravity problem, it would have a profound impact on our understanding of spacetime. We would likely learn more about the nature of gravity, and we would also learn more about the origin and fate of the universe.

Another challenge facing physicists is to understand the nature of spacetime at the quantum level. Spacetime is thought to be quantized, meaning that it is made up of discrete units. However, we do not yet know how to describe these units mathematically.

If physicists are able to understand the nature of spacetime at the quantum level, it would have a profound impact on our understanding of the universe. We would likely learn more about the nature of reality, and we would also learn more about the possibility of time travel and other exotic phenomena.

The evolution of our understanding of spacetime is likely to be a long and challenging process. However, it is a process that is essential to our understanding of the universe. As we continue to learn more about spacetime, we are likely to gain a deeper understanding of the cosmos and our place in it.

Here are some of the possible future developments in our understanding of spacetime:

- We may develop a theory of quantum gravity that reconciles the laws of general relativity with the laws of quantum mechanics.
- We may learn more about the nature of spacetime at the quantum level.
- We may discover new phenomena that are related to spacetime, such as time travel or wormholes.
- We may learn more about the origin and fate of the universe.

These are just a few of the possible future developments in our understanding of spacetime. As we continue to learn more about the universe, we are likely to make new discoveries that will change our understanding of spacetime.

The Implications Of Spacetime For Our Future As A Species

INTRODUCTION

The implications of spacetime for our future as a species are vast and profound. Our understanding of spacetime has the potential to revolutionize our way of life, and it could even determine whether or not we survive as a species.

Here are some of the potential implications of spacetime for our future:

- Space travel: Our understanding of spacetime could lead to new ways of traveling through space. We could develop new technologies that allow us to travel faster than the speed of light, or we could find ways to create wormholes that would allow us to travel through space instantaneously. This would open up the possibility of exploring other planets and galaxies, and it could even lead to the colonization of other worlds.

- Time travel: Our understanding of spacetime could also lead to the development of time travel. This would allow us to travel back in time and change the past, or it could allow us to travel to the future and see what the world will be like. Time travel is a controversial topic, but it is one of the most fascinating implications of spacetime.

- The nature of reality: Our understanding of spacetime could also lead to new insights into the nature of reality. We could learn more about the relationship between space and time, and we could also learn more about the origin and fate of the universe. This could have a profound impact on our understanding of ourselves and our place in the universe.

The implications of spacetime for our future are vast and profound. It is impossible to say for sure what the future holds, but our understanding of spacetime could play a major role in shaping it.

Here are some of the challenges that we may face in the future as a result of our understanding of spacetime:

- The possibility of time travel: If time travel is possible, it could pose a number of challenges. For example, it could lead to paradoxes, or it could allow people to change the past in ways that have negative consequences.
- The nature of reality: Our understanding of spacetime could also lead to new insights into the nature of reality. This could be unsettling for some people, as it could challenge their beliefs about the world.
- The future of the universe: Our understanding of spacetime could also give us insights into the future of the universe. For example, we could learn whether or not the universe is expanding forever, or we could learn whether or not there will be a Big Crunch.

These are just some of the challenges that we may face in the future as a result of our understanding of spacetime. It is important to be aware of these challenges so that we can be prepared for them.

INTRODUCTION

Spacetime is one of the most fundamental concepts in physics. It is the fabric of the universe, and it is what allows us to exist. Spacetime is also a mystery, and we do not yet fully understand it. The wonder and mystery of spacetime is something that has fascinated people for centuries. Philosophers, scientists, and artists have all pondered the nature of spacetime, and they have all come up with different interpretations.

One of the most beautiful things about spacetime is that it is a physical manifestation of the interconnectedness of all things. Spacetime is not just a collection of separate objects and events. It is a single, unified field that everything exists within.

This interconnectedness is also one of the things that makes spacetime so mysterious. We do not yet fully understand how everything in the universe is connected to everything else. However, we are beginning to learn more about the nature of spacetime, and we are likely to learn even more in the future.

WE ARE MORE LIKELY TO LEARN ABOUT OURSELVES

As we learn more about spacetime, we are also likely to learn more about ourselves and our place in the universe. Spacetime is the foundation of our reality, and it is what gives our lives meaning. The wonder and mystery of spacetime is something that we should all cherish. It is a reminder of the beauty and complexity of the universe, and it is a challenge for us to continue to learn and grow. Here are some of the things that we can do to reflect on the wonder and mystery of spacetime:

- Study the science of spacetime. There are many books, articles, and websites that can teach us about the science of spacetime. By studying science, we can gain a deeper understanding of this fundamental concept.

- Contemplate the philosophical implications of spacetime. Spacetime has profound philosophical implications. For example, it raises questions about the nature of reality and the relationship between space and time. By contemplating these implications, we can gain a deeper understanding of ourselves and our place in the universe.

- Create art that expresses the wonder and mystery of spacetime. Artists have long been fascinated by spacetime. They have created paintings, sculptures, and music that express the beauty and complexity of this fundamental concept. By creating art, we can share our own unique understanding of spacetime with others.

The wonder and mystery of spacetime is something that we should all appreciate. It is a reminder of the beauty and complexity of the universe, and it is a challenge for us to continue to learn and grow.

****

Spacetime In The Quantum World

This chapter would explore the intersection of spacetime and quantum mechanics, a field of physics that deals with the behavior of matter and energy at the atomic and subatomic levels. It would discuss the implications of spacetime for quantum gravity, a theory that seeks to unify the laws of general relativity (which govern the behavior of gravity on a large scale) with the laws of quantum mechanics (which govern the behavior of matter and energy on a small scale). Finally, it would explore the search for a unified theory of spacetime and quantum mechanics, a quest that has yet to be successful but that holds the promise of revolutionizing our understanding of the universe.

****

INTRODUCTION

Spacetime and quantum mechanics are two of the most fundamental concepts in physics. Spacetime is the fabric of the universe, and quantum mechanics is the theory of the very small. However, these two theories seem to be incompatible with each other.

Spacetime is described by general relativity, which is a theory of gravity that treats space and time as a single entity. Quantum mechanics, on the other hand, is a theory of the very small that treats space and time as separate entities.

This incompatibility is known as the quantum gravity problem. It is one of the most important unsolved problems in physics. There are a number of different approaches to solving the quantum gravity problem.

One approach is to develop a theory of quantum gravity that combines the principles of general relativity and quantum mechanics. Another approach is to develop a theory of quantum gravity that does not require the concept of spacetime.

The intersection of spacetime and quantum mechanics is a complex and challenging area of research. However, it is an area of research that is essential to our understanding of the universe.

Here are some of the possible implications of a unified theory of spacetime and quantum mechanics:

- A better understanding of gravity. A unified theory of spacetime and quantum mechanics could help us to better understand gravity. This could lead to new insights into the nature of black holes, neutron stars, and other astronomical objects.

- A better understanding of the universe. A unified theory of spacetime and quantum mechanics could help us to better understand the universe. This could lead to new insights into the origin and fate of the universe.

- New technologies. A unified theory of spacetime and quantum mechanics could lead to the development of new technologies. For example, it could lead to the development of new ways of traveling through space or new ways of manipulating matter.

The intersection of spacetime and quantum mechanics is a vast and fascinating area of research. It is an area of research that is likely to lead to new insights into the universe and new technologies.

Here are some of the challenges that we face in trying to understand the intersection of spacetime and quantum mechanics:

- The difficulty of combining general relativity and quantum mechanics. General relativity is a classical theory, while quantum mechanics is a quantum theory. These two theories are very different, and it is difficult to combine them.
- The lack of experimental evidence. There is no direct experimental evidence for a unified theory of spacetime and quantum mechanics. This makes it difficult to test and develop these theories.
- The complexity of the problem. The intersection of spacetime and quantum mechanics is a complex and challenging problem. It is a problem that has not yet been solved.

Despite these challenges, the intersection of spacetime and quantum mechanics is an area of research that is full of promise. It is an area of research that is likely to lead to new insights into the universe and new technologies.

INTRODUCTION

Spacetime is the fabric of the universe, and it is the stage on which the laws of physics play out. Quantum gravity is a theory that attempts to unify the laws of general relativity, which describes gravity on a large scale, with the laws of quantum mechanics, which describe the behavior of matter and energy on a small scale.

The implications of spacetime for quantum gravity are vast and profound. If we can understand how spacetime works at the quantum level, we will be able to understand the nature of gravity at the quantum level. This would have a profound impact on our understanding of the universe.

Here are some of the implications of spacetime for quantum gravity:

- The nature of gravity. Quantum gravity could help us to understand the nature of gravity. For example, it could help us to understand why gravity is so weak compared to the other forces of nature.

- The origin of the universe. Quantum gravity could help us to understand the origin of the universe. For example, it could help us to understand how the universe came into existence and how it has evolved over time.

- The future of the universe. Quantum gravity could help us to understand the future of the universe. For example, it could help us to understand whether the universe will continue to expand forever or whether it will eventually collapse in on itself.

The implications of spacetime for quantum gravity are vast and profound. It is an area of research that is likely to lead to new insights into the universe and new technologies.

Here are some of the challenges that we face in trying to understand the implications of spacetime for quantum gravity:

- The difficulty of combining general relativity and quantum mechanics. General relativity is a classical theory, while quantum mechanics is a quantum theory. These two theories are very different, and it is difficult to combine them.

- The lack of experimental evidence. There is no direct experimental evidence for quantum gravity. This makes it difficult to test and develop these theories.

- The complexity of the problem. The implications of spacetime for quantum gravity is a complex and challenging problem. It is a problem that has not yet been solved.

Despite these challenges, the implications of spacetime for quantum gravity is an area of research that is full of promise. It is an area of research that is likely to lead to new insights into the universe and new technologies.

The Search For A Unified Theory Of Spacetime & Qquantum Mechanics

INTRODUCTION

Spacetime and quantum mechanics are two of the most fundamental concepts in physics. Spacetime is the fabric of the universe, and quantum mechanics is the theory of the very small. However, these two theories seem to be incompatible with each other.

Spacetime is described by general relativity, which is a theory of gravity that treats space and time as a single entity. Quantum mechanics, on the other hand, is a theory of the very small that treats space and time as separate entities.

This incompatibility is known as the quantum gravity problem. It is one of the most important unsolved problems in physics.

There have been many attempts to find a unified theory of spacetime and quantum mechanics. Some of the most promising approaches include:

- String theory: String theory is a theory that proposes that all particles are made up of tiny strings. These strings vibrate in different ways, and the different vibrations give rise to different particles. String theory is a promising approach to quantum gravity because it treats space and time as a single entity.

- Loop quantum gravity: Loop quantum gravity is a theory that proposes that spacetime is made up of tiny loops. These loops interact with each other in a way that gives rise to gravity. Loop quantum gravity is a promising approach to quantum gravity because it treats space and time as discrete entities.
- M-theory: M-theory is a theory that combines string theory and loop quantum gravity. It is a very complex theory, but it is a promising approach to quantum gravity because it may be able to unify all of the forces of nature.

The search for a unified theory of spacetime and quantum mechanics is a long and challenging journey. However, it is a journey that is essential to our understanding of the universe.

Here are some of the challenges that we face in the search for a unified theory of spacetime and quantum mechanics:

- The difficulty of combining general relativity and quantum mechanics. General relativity is a classical theory, while quantum mechanics is a quantum theory. These two theories are very different, and it is difficult to combine them.

- The lack of experimental evidence. There is no direct experimental evidence for a unified theory of spacetime and quantum mechanics. This makes it difficult to test and develop these theories.
- The complexity of the problem. The search for a unified theory of spacetime and quantum mechanics is a complex and challenging problem. It is a problem that has not yet been solved.

Despite these challenges, the search for a unified theory of spacetime and quantum mechanics is an area of research that is full of promise. It is an area of research that is likely to lead to new insights into the universe and new technologies.

****

Thank You For Joining Me!

We have journeyed through the fabric of spacetime, from its origins to its future. We have discussed the history of spacetime, the different theories of spacetime, and the mysteries that still surround spacetime. We have also explored the implications of spacetime for our understanding of the universe and our future as a species.

Spacetime is a complex and mysterious place. There is still much that we do not understand about it. However, the more we learn about spacetime, the more we learn about the universe itself. And as we learn more about the universe, we are learning more about ourselves.

The Fabric of Space & Time is a journey of discovery. It is a journey that has taken us to the very heart of the universe. It is a journey that has changed the way we see the world.

We have come to the end of our journey, but our exploration of spacetime is just beginning. There is still much that we do not know, and there are still many mysteries to be solved. However, we are confident that, with the help of science and technology, we will continue to learn more about spacetime and its mysteries.

We hope that you have enjoyed our journey through the fabric of spacetime. We hope that you have learned something new and that you have been inspired to continue your own exploration of the universe.

Thank you for joining me!

Leave Reviews

You can leave your reviews about the book on my blog @ https://sciencemaverick.blogspot.com Also. you can email me @ scienceoftheweirdandwonderful@gmail.com

Copyright © Policy